Cipariu Adriana-Ioana
Tita Mihaela-Adriana
Tita Ovidiu

Traceability study of the hard paste cheeses: The Romano Cheese

AF 153517

Table of Contents

1.INTRODUCTION..3

2.HARD PASTE CHEESES...5

➤Cheddar cheese...6
➤Parmesan cheese..8
➤Swiss Cheese (Emmental).. 9
➤Romano cheese...11

3.QUALITY ASSURANCE OF THE ROMANO CHEESE............................18

Milk quality as raw material..19
Pathogenic agents in raw milk..19
Potentially present chemicals in milk..20
Physical risks in milk..21
Scheme of phase process control for Romano cheese..................................21
Applying the HACCP system in the cheesemaking process............................24
HACCP principle application in the Romano cheesemaking process..................25

4.TRACEABILITY STUDY OF ROMANO CHEESE MAKING.....................35

General aspects of traceability...35
Usefulness of traceability systems..37
Traceability system features...37
Usage of traceability systems in the Romano cheese making process.................38
Database operation..48
Conclusions..50
BIBLIOGRAPHY..51

TRACEABILITY STUDY OF THE HARD PASTE CHEESES: THE ROMANO CHEESE

SCIENTIFIC COORDINATOR:

Profesor Univ. Dr. Ing. MIHAELA-ADRIANA TIȚA

Profesor Univ. Dr. Ing. OVIDIU TIȚA

STUDENT:

Cipariu Adriana-Ioana

Impressum / Imprint

Bibliografische Information der Deutschen Nationalbibliothek: Die Deutsche Nationalbibliothek verzeichnet diese Publikation in der Deutschen Nationalbibliografie; detaillierte bibliografische Daten sind im Internet über http://dnb.d-nb.de abrufbar.
Alle in diesem Buch genannten Marken und Produktnamen unterliegen warenzeichen-, marken- oder patentrechtlichem Schutz bzw. sind Warenzeichen oder eingetragene Warenzeichen der jeweiligen Inhaber. Die Wiedergabe von Marken, Produktnamen, Gebrauchsnamen, Handelsnamen, Warenbezeichnungen u.s.w. in diesem Werk berechtigt auch ohne besondere Kennzeichnung nicht zu der Annahme, dass solche Namen im Sinne der Warenzeichen- und Markenschutzgesetzgebung als frei zu betrachten wären und daher von jedermann benutzt werden dürften.

Bibliographic information published by the Deutsche Nationalbibliothek: The Deutsche Nationalbibliothek lists this publication in the Deutsche Nationalbibliografie; detailed bibliographic data are available in the Internet at http://dnb.d-nb.de.
Any brand names and product names mentioned in this book are subject to trademark, brand or patent protection and are trademarks or registered trademarks of their respective holders. The use of brand names, product names, common names, trade names, product descriptions etc. even without a particular marking in this work is in no way to be construed to mean that such names may be regarded as unrestricted in respect of trademark and brand protection legislation and could thus be used by anyone.

Coverbild / Cover image: www.ingimage.com

Verlag / Publisher:
LAP LAMBERT Academic Publishing
ist ein Imprint der / is a trademark of
OmniScriptum GmbH & Co. KG
Heinrich-Böcking-Str. 6-8, 66121 Saarbrücken, Deutschland / Germany
Email: info@lap-publishing.com

Herstellung: siehe letzte Seite /
Printed at: see last page
ISBN: 978-3-659-80191-4

Zugl. / Approved by: Sibiu, University of Sibiu, 2015

Cipariu Adriana-Ioana
Tita Mihaela-Adriana
Tita Ovidiu

Traceability study of the hard paste cheeses: The Romano Cheese

LAP LAMBERT Academic Publishing

1. INTRODUCTION

People have consumed milk and learned to make cheese from it from ancient times. In our country, the beginning of cheesemaking must be quite long, considering the fact that our ancestors, the Thracians and the Dacians had as main occupation the breeding of livestock, mainly cattle and sheep. [3]

The begining of cheesemaking is lost in the mists of time, being one of the oldest manufactured and used food in human nutrition. Cheese is also the oldest method of milk preservation, initially made from empirical recipes, some of which are still valid today.

In the French concept (France being the most renowned country for cheesemaking; here they process up to 500 varieties of cheese), the word cheese is reserved for the fermented or unfermented product, obtained by coagulating milk, cream, écrémé or mixed milk, followed by whey draining with a 23% minimum dry matter.

There are currently more than 1,000 types of cheese worldwide, which differ depending on the coagulation, the processing and draining method, as well as the maturation process, resulting in the defining characteristics related to taste, aroma, texture and form.

From the wide range of dairy products, cheese represents the most rich and varied group, compresing a high nutritional value per unit mass, much higher compared to milk itself, or with other acid dairy products, being rightly considered a protein food of the highest quality.

In the last 3-4 decades, especially in the developed countries, an increased attention has been given to the production of varieties that better fit the quality standards and directions stipulated by legislation. Thus, cheese varieties with designation of a particular origin, have appeared on primary markets, which is to guarantee the consumer a quality product as for these types, bearing a certain trade mark, a very strict control throughout the flow process is established [9]

Cheese is defined as the fresh or matured product, obtained by draining the whey, after the coagulation of milk, cream, skimmed milk, buttermilk or by mixing one or all of these products.

Cheese products are foods with a high nutritional value due to the content in protein, fat, minerals and vitamins, nutrients of very good quality and with high bioavailability. [10]

Cheese, as a result of applied biotechnology, is one of the most complex and dynamic foods. Each piece can be considered a bioreactor, in which numerous and complicated reactions are produced, the final outcome being a product with specific sensory characteristics.

Cheese is an excellent food due to its high nutritional value, good digestibility and the pleasure its consumption generates. The nutritional interest for this food resides mainly in its composition, which includes, in a relatively large proportion, proteins with high biological value, calcium, phosphorus and some vitamins, especially A and D. [6]

The chemical composition of cheese and implicitly its nutritional contribution, widely varies depending on the used milk (animal species, source, lactation period), the technology, the variety of cheese, the season in which it is produced, and even the standards according to which it is manufactured. [10]

The nutritional and biological value of cheese depends on its chemical composition, which in its turn is influenced by the assortment. The nutritional characteristics of cheese essentially consists of the following:

- Very high digestibility, 97-98% compared to 90% for other products of animal origin;
- High biological value, 84% compared to 73% for meat and 44% for beans;
- High nutrient composition, 18-36% proteins, 16-38% fat, 2.7 to 5.0% mineral salts, as well as high content of calcium, phosphorus, sodium, manganese, magnesium and vitamins (A, B1, B2, B6, D, PP, K);
- High nutritional value (100g cheese has a nutitional power equivalent to 700 g veal, 400 g pork, 500 g fish, 8-9 eggs, 1400g apples);
- High energy value 3800 kcal / kg;
- High hygienic and dietetic value;
- High appetite value.

Due to its nutritional qualities, cheese plays a crucial role, thus ensuring the nutritional requirements of human nature. It can be used both as a dietetic food as well as an appetizer or dessert, this depending from country to country, according to tradition and preferences. [9]

Due to its specificity, making cheese has long been linked to traditional and artisanal methods of preparation. As the transit was made from small series to industrial production, the need for mechanisation and automation of manufacturing processes appeared.

Therefore, the aim is to reduce labour, which in the cheese making process represents a heavy workload, as well as obtaining products of constant and uniform quality.

For semi-hard and hard cheeses, mechanization was made a while ago, when the mechanical vat was introduced. [3]

2. HARD PASTE CHEESES

Cheese represents the most rich and varied dairy group, its preservation being ensured by lactic acidification, low water content, with and without added salt.

Currently, depending on the nature of the milk used as raw material, and the applied technology, with many variable parameters, a varied assortment of cheeses can be obtained. This diversity makes it difficult to provide an accurate classification of cheeses, perfectly separated into groups, specifying an exact framing for any assortment in any of these groups.

The name of most types of cheese is linked to the place or locality where they were first obtained.

Cheese classification is based on the following criteria:

- Milk type;

- Fat content of the cheese;

- Paste consistency;

- The manufacturing process. [3]

Given the paste consistency, cheeses are classified as follows:

- semi-hard, with the following characteristics:

- Curdling temperature 30-34 °C;

- Higher degree of grinding, as for soft cheeses;

- Apply a second heating to 35-45 °C;

- Strong pressure;

- Maturation time of 1-2 months;

- Well-formed rind;

- Produced in a 2-5 kg format .

Assortment examples: Trappista cheese, Dutch cheese, Harghita cheese.

- hard, with the following characteristics:

    - Short duration for the curdling process;

    - High degree of curd grinding;

    - Apply a second heating to 50-58 °C;

    - Strong pressure;

    - Long maturation time;

    - Low water content;

    - Long-preservation.

Examples: Swiss cheese, Muresana cheese, Cheddar cheese, Parmesan, Roomano. [10]

> **Cheddar cheese**

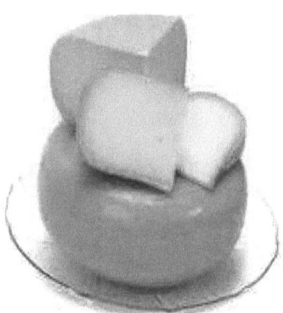 

Figure no.2.1. Cheddar cheese

The story of the Cheddar cheese can be traced to Roman times, who then introduced cheese to Great Britain. It has since been emulated around the world, especially in Canada, Australia and New Zealand, where it is mostly made in blocks rather than cylinders. But only the ones produced with milk coming from the cows grazing on the green hills of England really deserve the name of Cheddar cheese.

For generations, Cheddar cheese has been part of the English diet, used for sandwiches, snacks, quick lunches, or displayed on huge platters of cheese with pickled nuts and fresh bread. It is also lovely with sauces, melted over baked potatoes, or grated over numerous vegetable dishes and steaks. It is best served with a Merlot or Pinot Noir. [7]

Chedarring is the specific technological phase for this cheese assortment. The purpose of this operation is to allow acidity and to remove as much whey as possible. During the cheddaring process, a larger amount of lactic acid is formed, thus inducing the accumulation of monocalcium para-casein; green cheese modifies its consistency, becoming softer, it stretches and unfolds in thin layers. The cheddar procees is considered complete when the whey reaches an acidity of 60-70°T whey, green cheese acidity is about 220°T.

Cheddar cheese is produced in a two cylinder format or is parallelepiped shaped:

- type I- diameter of 36-39 cm, height 28-35 cm and weight of 30-35 kg;

- type II diameter of 36-39 cm, height 15-18 cm and weight of 15-18 kg;

- the rectangular shape has the following dimensions: length 35-36 cm, width 27-28 cm, height 17-18 cm.

Characteristics of the Cheddar cheese

Organoleptic characteristics

The exterior surface is smooth, without cracks, with a continuous and adherent layer of waxed cloth or plastic material. In section, the compact mass has no fermentation „eyes" and may present elongated holes due to the pressing process.

The yellowish-beige uniform core takes the form of a fine paste, slightly elastic.

Pleasant taste, with a specific nutty flavour. [3]

Chemical characteristics

Table no. 2.1 Chemical characteristics of the Cheddar cheese

Characteristics	Admisability conditions
Water % max	40
Fat% DM min	48
Sodium chloride, %	1,5-2,5

> **Parmesan cheese**

Figure no.2.2. Parmesan cheese

This assortment is part of a cheese group called scraping cheeses, which can be preserved for several years and has a very low water content (under 30%). These cheeses are not used for direct consumption, but as addition to different foods and dishes to enhance their taste.

Its original homeland is Italy- Fromaggio di Grana. There are several types, depending on the region of Parma, named Cacio Parmigiano or simply Parmesan, a very well liked assortment that has known a great extension.

The use of milk with a high degree of maturation, with an acidity of 21-23°T within the production process is typical for this cheese assortment.

Parmesan cheese shelf life is 15-20 years at a temperature of 10-12°C and low humidity. The older the cheese, the higher the price, therefore it is customary to mark the manufacturing date on it.

Parmesan cheese is presented in a cylindrical shape with the two bases completely flat and almost straight margins with slightly rounded edges. The diameter varies between 35 and 65 cm, 18 to 20 cm high and weights 25-30 kg

Organoleptic characteristics of the Parmesan cheese

The exterior presents itself as a completely smooth, hard, dark brown rind. Inside, the pulp is uniform and yellowish, and presents rare small eyes.

Consistency is very hard, fine-grained and presents radial cleaving; nevertheless a certain characteristic elasticity is present, low friable so that it can be easily grated.

It also presents a characteristic odor and pleasant, aromatic taste depending on the its age.

Chemical characteristics

Table no.2.2. Chemical characteristics of the Parmesan cheese

Characteristics	Admisability conditions
Water, %	25-35
Fat % DM	37-40
Sodium chloride, %	1,5-2,2

[4]

➤ **Swiss Cheese (Emmental)**

Figure no.2.3. Swiss cheese

The origins of the Emmental cheese, one of the great classics, can be traced back to 1293 [7].

The homeland of the Swiss cheese is Switzerland, where it is called Emmental, originating from the area of the Emme Valley, where even today, high quality cheese is produced.

Due to its particular organoleptic characteristics, the making of this cheese has spread in most countries with a developed dairy industry such as Austria, France, Finland, Sweden.

Obtaining quality Emmental cheese depends primarily on the raw material used; the milk should be fresh, sanitized and of superior quality; milk collected from cows at the beginning and at the end of the lactation period cannot be used, neither that coming from cows suffering from any disease.

Emmental is part of the fine cheese category, with a rather complicated manufacturing process which requires great skill and experience from the part of the cheese maker leading the entire production process. Because the success of this process involves appropriate weather conditions, these cheeses are produced in mountainous regions.

Swiss cheese is presented as cylindrical wheels, with a slightly convex lateral surface, with a diameter of 70-80 cm, 13-18 cm high and weights between 60 and 100 kg.

Organoleptic characteristics of the Swiss cheese

The outer rind, of a pale yellow to dark yellow colour is smooth, elastic, moderately thick, resistant, and slightly greasy to the touch, and present traces of cheesecloth.

The inside is a compact paste, homogenous, clean, of a yellowish colour, with round cherry size or "bull's eye" size fermentation eyes. The inner surface of the eyes is smooth, with a bright appearance. For high quality Swiss cheese, the distribution of the fermentation eyes is almost regular; the sample removed with a probe presents two or maximum three eyes. Consistency is elastic, smooth, but not crumbly and melts in the mouth. The smell-taste are pleasant, aromatic, smooth, slightly sweet, green walnutty like. [3]

Chemical characteristics

Table no.2.3. Chemical characteristics of the Swiss cheese

Characteristics	Admisability conditions
Water, % max	42
Fat % DM min	45
Sodium chloride, % max	1,5

> **Romano cheese**

Figure no.2.4. Romano cheese

The Romano cheese is a typical Italian one, traditionally produced in Lazio and lately also in Sardinia.

The cheese paste is compact and easily pierced with holes and its color varies from white to pale yellow, more or less intense, depending on the technical production conditions. The taste is flavored, slightly spicy and savory for a table cheese, intensely spicy for a grated cheese.

The following table shows the nutritional characteristics of 100 grams Romano cheese:

Table no. 2.4 Nutritional value for 100g Romano cheese

Energetic value	390,918 kcal/1546 kJ
Proteins	24,6 g
Lipids	31,87 g
Carbohydrates	< 1 g
Calcium	932 mg

[13, 14]

Romano cheese comes in the form of cylindrical pieces with slightly rounded edges and presents the characteristics in the table below.

Table no.2.5. Characteristics of the Romano cheese

Characteristics	Admisability conditions
Weight, kg	6
Height, cm	9-12
Diameter, cm	27-30

Organoleptic properties

Table no. 2.6 Organoleptic characteristics of the Romano cheese

Characteristics	Admisability conditions
External appearance	Yellowish-white rind without mould The surface may be paraffined
Interior appearance	Yellowish-white paste, evenly spread throughout its mass, with small fermentation holes and small pressing holes
Core consistency	Hard, for grating
Smell and taste	Spicy, salty, pleasant, typical for cheese produced from fermented cow's milk

Chemical properties

Table no. 2.7 Chemical characteristics of the Romano cheese

Characteristics	Admisability conditions	Analysis methods
Fat % DM min	34	STAS 6352-61
Water % max	38	STAS 6344-68
Sodium chloride % max	9	STAS 6354-70

Bacteriological properties

Pathogenic germs - absent

Quality control rules

The quality control of the Romano cheese is made on batches of max. 1000 kg.

- Marking and packaging control

- Organoleptic examination

- Chemical analysis.

A microbiological examination will be performed on request.

For the packaging and marking control, as well as for the bacteriological examination, a number of packages are taken at random from the batch, according to the below table:

Table no. 2.8 Bacteriological examination

Number of packages from the batch	Number of packages to be analysed
Up to 20	2
From 21-40	3
From 41-60	4

From 61-100	5

After checking the packaging and marking as above, a randomly taken piece of cheese is cut in radial slices of about 200 g and examined organoleptically.

For the chemical examination approx. 100 g of cheese are taken from one or two pieces of the open packages.

The sample is placed in a clean and tightly closed glass jar, with a polished or waxed stopper, which is sealed and labeled. The label covers the following specifications:

- Name and address of producer,
- Product name and type,
- Internal standard 1395 -67,
- Date and place of the sample taking,
- The name and signature of the persons who took the sample.

Samples for microbiological analysis are taken separately, according to STAS 6349-71.

Analysis methods

Preparation of samples for physico-chemical analysis according to STAS 6343-71. Organoleptic examination, according to STAS 6345-71. Microbiological analysis, according to STAS 6349-71.

Determination of fat, water and sodium chloride, according to the STAS mentioned above.

Packaging and Marking

Each piece of Romano cheese is packed in metallic paper and wrapping paper.

Thus packed, the cheese is placed in wooden boxes (4 to 6 pieces in a box).

Each piece is marked with a label or a metal sticker with the following specifications:

- producer's name or trademark,
- product name and type,
- price per kilo,
- production date.

The boxes are marked by stamping with the following specifications:

- producer's trademarks and address,

- name and type of product,

- package order number,

- net weight,

- country,

- delivery date.

Storage and transport

Ready fermented Romano cheese is stored in cold, clean, disinfected without foreign smells, well ventilated stores, with a temperature of 1-5°C and 85-90% relative humidity.

The Romano cheese wheels are stored only on shelves, the pieces put on one of their bases. The pieces cannot overlap.

Transport vehicles will be covered, clean, dry and free from foreign smells with moderate temperatures. In warm weather, transportation will be made in max. 12 hours.

Provided that storage and transport conditions mentioned above are met, the warranty period is 8 months from the production date.

Each consignment will be accompanied by a quality certificate. [11]

Romano cheese defects

Cheese is characterized by certain organoleptic and physico-chemical properties typical for each assortment, characteristics that are governed by internal rules and standards. However, sometimes, the products do not meet these requirements due to the failure of compliance with the conditions imposed by the production process or the use of faulty raw materials. In this case, cheese will show serious or less serious defects, which can sometimes make it unfit for consumption. Recognizing these defects and their root-causes is of particular importance for all cheese makers, so that they can prevent them by adopting appropriate actions. [4]

Defects that may result are due to:

- inadequate milk quality as it may be infected with harmful microorganisms, as a result of some fodder or poor animal health;
- failure of compliance with the manufacturing technology;
- action of pests (insects, rodents). [2]

Defects that may result in the Romano cheese are the following:

15

Uneven pieces, with large inside holes due to high acidity of the processed milk which does not allow a uniform interfusion of the curd grain, a far too rapid maturation process also contributing to this.

A mouldy rind may appear on surfaces where paraffin has removed itself due to a poor or premature paraffining, high temperature and humidity within the maturation phase also contributing to the mould formation.

Coarse composition, the paste large grains due to the use of high acidity milk in the production process.

A cracked rind is the result of an insufficient pressing or low humidity within the pre-maturing rooms (surface drying).

Crumbly paste appears when milk with high acidity is used or when the prescribed acidity during the curd processing is exceeded.

To control this defect, it is recommended that milk with normal acidity be used and excessive acidity growth within the production process be prevented, especially in summer.

Hard paste defect appears when cheese is strongly pressed, having a low water content and low acidity. [10]

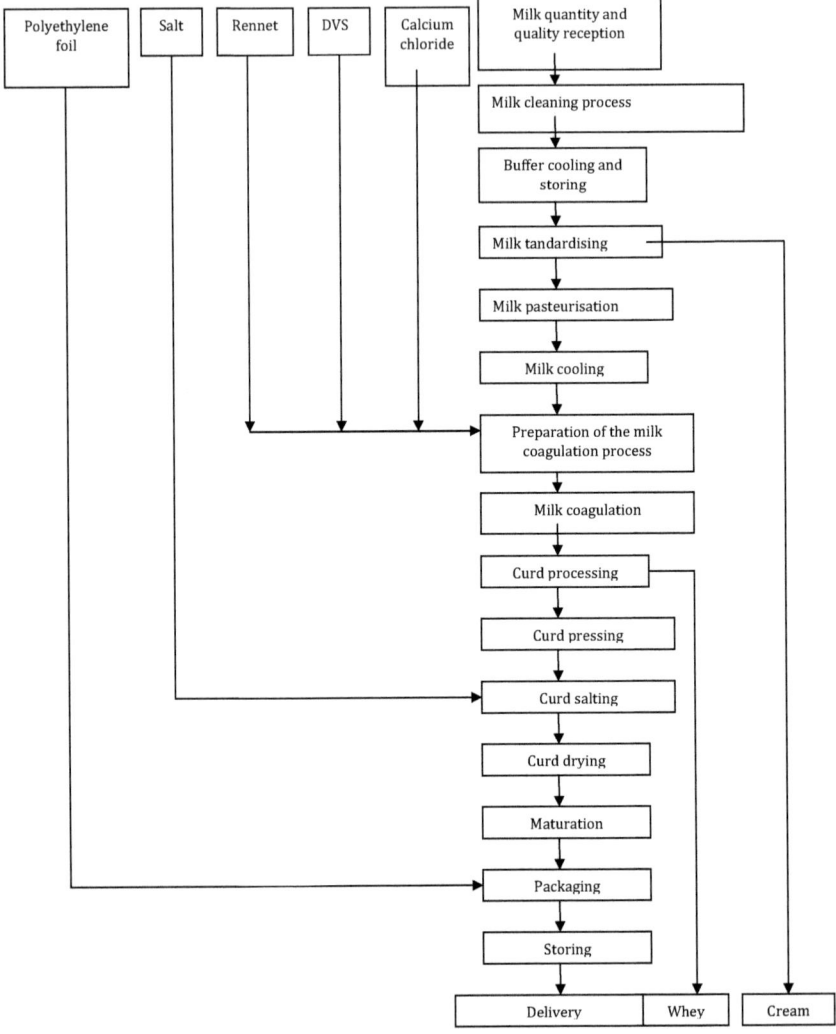

Figure no. 2.5 Tehnological scheme of the Romano cheesemaking process

3. QUALITY ASSURANCE OF THE ROMANO CHEESE

To assess cheese quality, a series of tests are carried out, thus allowing us to establish whether those products fall within the provisions of the standards or technical specifications and whether they can be consumed.

The control process implies:

- Packaging and labeling control - organoleptic examination
- Chemical and microbiological analysis

Packaging and marking check - cheese quality control begins by checking the packaging, marking and external appearance of a large number of packages or wheels according to the established standards, depending on the analyzed batch size and the product's characteristics.

The organoleptic characteristics of the product are determined by the specified instruction standards, after the examined samples are brought to a temperature of 15 ... 20 °C.

Obtaining high, uniform and constant quality products is ensured by carefully monitoring the technological process during the cheese making operation, in full compliance with the recommended parameters.

It is necessary to obtain different batches of cheese in order to determine the main parameters of the manufacturing process, which are part of a specification sheet. It is thus possible to analyse the manufacturing process, either to take measures in order to prevent defects, or to understand the causes of these defects.

Specification sheets can be detailed or simplified, covering only the main parameters of the manufacturing process. These specification sheets are of particular importance, especially for the production of those cheese assortments where drawing, consistency and gas-producing defects occur. [3]

Cheese quality depends on the quality of the raw and auxiliary materials used in its manufacturing process.

18

➤ The milk used in the cheese making process must meet the following requirements: it must come from healthy animals, rationally fed with quality forage so that foreign taste and smell are not transmitted to the milk;

➤ it must be microbiologically compliant (low number of total germs and mainly sporogenous anaerobic bacteria Clostridium-type or butyric bacteria that are late cheese gas-producing agents);

➤ cow milk must have an acidity of 16-20°T

➤ milk obtained in the first 8-10 days after calving and in the last 10-15 days of lactation must not be used;

➤ milk must not contain antibiotics and antiseptics that have inhibiting action on the microorganisms that ensure maturation.

Quality control of milk used in the cheese production implies:

- Sensory analysis: colour, taste, smell, appearance;

- Physico-chemical analysis: degree of contamination, density, fat content, acidity, protein titre;

- Microbiological analysis: reductase testing, fermentation testing, rennet coagulation testing, establishing the spore-forming bacteria load. [10]

Milk quality as raw material

• **Pathogenic agents in raw milk**

Raw milk contains numerous pathogenic agents and may be a source for diseases as brucellosis, tuberculosis or other zoonoses, as well as for numerous other food poisoning related ones.

Milk contamination can occur in various ways, but the most important is the contamination with faeces and it mainly takes place during the milking process. Even under very good milking conditions by maintaining strict hygiene rules, it is impossible to totally eliminate the faecal contamination risks.

Preventing excessive contamination and recontamination can be achieved by adhering to good working practices (GMP) including good hygienic practices (GHP), the HACCP supplier plan and directly or indirectly monitoring the milking process by performing periodic audits. [6]

19

- **Potentially present chemicals in milk**

A number of chemicals may end up into milk by direct transfer from the animal, forage or the environment and may affect the cheese consumers. **Growth hormones, antibiotics and other antimicrobial medicine.** No long-term effects are known (especially in children) related to the presence of residues of such substances used in the animal breeding and treatment.

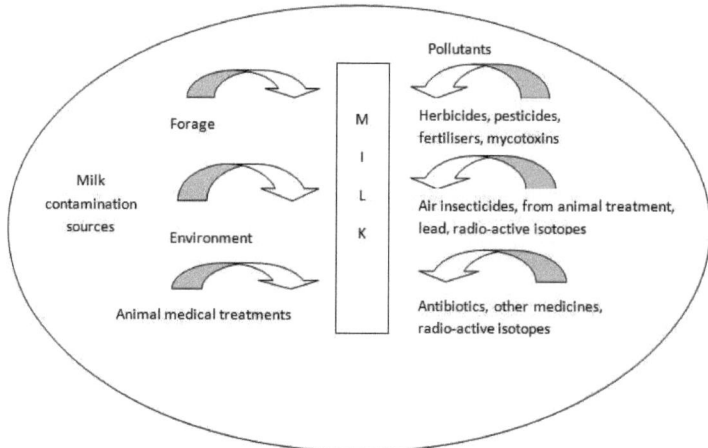

Figure no. 3.1 Sources of milk chemical contamination

Mycotoxins. The presence of mycotoxins in milk is a result of indirect sources or mould infected toxigenic forage. The most common risk is the presence of aflatoxin M1.

Radio-active substances. The risk of radionuclide contamination is high because the contaminating substances present in the grazing land can rapidly transfer into the milk. It is well known the role of radioactive iodine in the development of thyroid cancer in children.

- **Physical risks in milk**

   The physical risk assessment (hairs, straws, wood chips, glass, pebbles and dust) proved to be of minor importance, being easily removed through filtration or centrifugation, but they are an important source of milk contamination with pathogenic germs. [6].

- **Scheme of phase process control for Romano cheese**

Table no. 3.1. Scheme of phase process control

No.	Name of thenological operation	Name of monitored product	Laboratory analysis	Value limitation
1.	Quality acceptance	Whole milk	Organoleptic	White-yellowish opaque liquid,normal consistency.
			Acidity	16-20°T, pH 6,3-6,9
			Density 20°	1,027-1,034 g/cm³
			Fat	3,5-3,9%
			Proteic titre	minimum 3,2%
			Dry substance	9-9,5%
			Reductase examination	Discolouration within 5h NTG= max. 300.000/cm³
			Microbiological examination	Coliform bacteria= max. 10/cm³
2.	Quantity acceptance	Whole milk	Quantity, kg	-
3.	Centrifugal cleaning	Whole milk	Contamination degree	2nd degree maximum

4.	Cooling and buffer storage	Cold milk	Temperature	4-6°C
5.	Standardisation	Standardised milk	Fat	3,4%
6.	Pasteurisation	Pasteurised milk	Phosphatase test	Yellow colouring
			Temperature	65°C
			Time	20 minutes
7.	Cooling	Cold	Temperature	40°C
8.	Milk preparation for the curdling-maturing process	Cold-matured, pasteurised milk	Quantity of $CaCl_2$, DVS	7,26 kg; 0,53 kg
			Temperature	32°C
			Acidity	max. 21°T
			Maturation time	15 min
9.	Coagulation	Curd	Rennet quantity	0,508 kg
			Acidity	21°T
			Temperature	40°C
			Time	15-18 minutes
10.	Curd processing	Cut curd	Dimension of curd grain	2-5 mm
			Time	55-60 minute
			Processing temperature	45°C
			Curd grain acidity	21°T
				50-60°T

		Whey	Acidity	
11.	Forming-pressing	Green cheese	Pressing time	approx 45 minutes
			Green cheese acidity	60-65°T
				63-65%
			Humidity	
				45°C
			Temperature	
12.	Dry salting	Green cheese	NaCl content	570,77 kg
			Humidity	58-62%
			Salting time	3 months
		Salt	Humidity	max. 0,15%
13.	Surface dry	Green cheese	Humidity	45%
14.	Maturation	Maturated cheese	Maturation degree	Fresh cheese
				min. 34%
			Fat/DM	
				250°T
			Total acidity	
				pH=6
			Active acidity	
				max. 38%
			Humidity	
				max. 9%
			NaCl content	
16.	Packaging	Polyethiylene foil	Microbiological package examination	NTG= max. 2/cm³
				Coliform bacteria= abs.

				/10 cm³
17.	Storage	Packed cheese	Temperature	4°C
			Air humidity	85-90%
18.	Delivery	Delivered cheese	Organoleptic	White-yellowish homogeneous paste, hard consistency, salty taste.
			Humidity	max. 38%
			Total acidity	250°T
			Fat/DM	min. 34%
			NaCl content	max. 9%
			Temperature	0-8°C

- **Applying the HACCP system in the cheesemaking process**

Statistics have revealed food related diseases due to pasteurized milk, milk powder, ice cream and acidic dairy products.

For a long time, cheeses were considered as a safe food, although even in developed countries, food poisoning caused by Salmonella spp., Listeria monocytogenes, Escherichia coli was registered.

Even if, cheese causes less food poisoning compared to other foods, others, such as the chemical and physical aspects related to safety consumption, must be kept under control. Given the high infection degree of raw milk with pathogenic germs, the possible toxic pollution, as well as microbiological and chemical contamination possibilities during the manufacturing process, the use of the HACCP system has become essential in the cheese making industry. [6]

The fundamental functions of the HACCP method are:
- Hazard analysis;

- Identification of critical points;
- Execution supervision;
- System effectiveness verification.

From the mentioned functions, derive 7 principles of the HACCP. [1]

- **HACCP principle application in the Romano cheesemaking process**

Principle 1. Conducting the hazard analysis

Under this principle, a systematic analysis of the food product, subject of the application, and its ingredients is made.

The purposes of the analysis are:

➢ Identifying the presence of pathogen microorganisms, parasites, chemicals or foreign bodies that could affect consumers' health.

➢ Including the product in a certain hazardous category

➢ Risk classification in a certain severity category, useful for the product and ingredient hazard assesment [1]

To identify hazards, the following two tools are used: the matrix and the ishikawa.

The matrix is a method based on 4 significance levels that define 4 classes of risk, and the classification in each of these classes is determined by the types of control measures to be taken, namely:

Risk class 1 - does not impose any control measure.

Risk class 2 – need for personnel awareness procedures.

Risk Class 3 - general control procedures are required.

Risk Class 4 - requires mandatory specific control procedures.

Table no. 3.2. Matrix

Gravity	Risk class		
High	3	4	4
Medium	2	3	4
Low	1	2	3
Frequency	Low	Medium	High

25

Table no. 3.3 Hazard analysis in the Romano cheese making process

Raw materials/Stage	Detection risks	Gravity	Frequency	Risk class	Control measurements
Rennet	Biological - Salmonella presence	High	Low	3	-compliance with transport and storage conditions; - supplier selection;
	Chimical - presence of antibiotics, disinfectants	Medium	Low	2	-application of good hygiene practices;
Pasteurisation	Biological -Salmonella presence	High	Low	3	-compliance with the pasteurisation parameters;
Coagulare	Chimical -chemical substances from machines	Medium	Low	2	- Compliance with hygiene norms
Cleaning	Biological -presence of coliform bacteria	Medium	Low	2	- usage of microbiological suitable drinking water
Packaging	Biological -packaging mould contamination	Medium	Low	2	-supllier selection; -appropriate storage of the packaging material.
Storage Delivery Transport	Biological -Yeast and mould contamination due to hygiene failure	High	Low	3	-appropriate storage and transport conditions; - compliance with hygiene norms.

The second tool used for hazard analysis is the ishikawa, also known as fishbone, the 5 "M" rule or cause-effect. It is based on identifying potential causes that induce non-conformities, biological, chemical or physical contaminations.

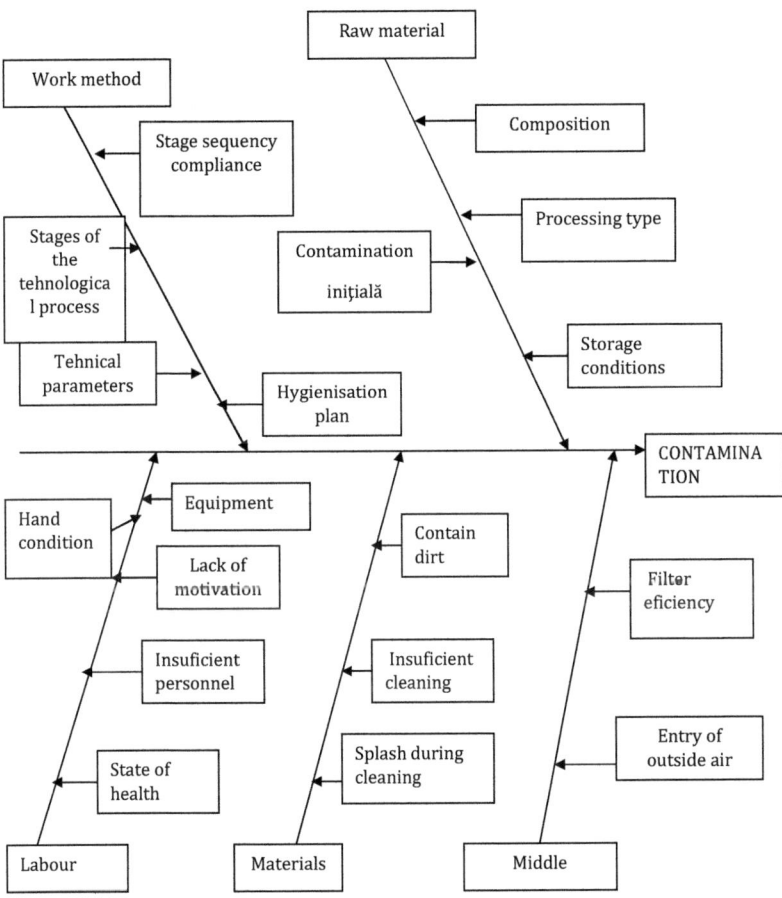

Figure no. 3.2. Rule five M for pasteurization

Principle 2. Determination of critical control points (CCP)

A critical control point is defined as any point or procedure of a system specialized in the production of foods in which the loss of control can result in endangering the health of consumers.

All identified hazards must be eliminated or reduced within a certain stage of the manufacturing cycle, from the cultivation / growing and harvesting of raw materials stage to product consumption. The critical control points can be located at any stage of the technological process where it is required and it is possible to control harzardous microorganisms or risks of any kind. [1]

Identification of critical control points is made based on the decision tree shown below.

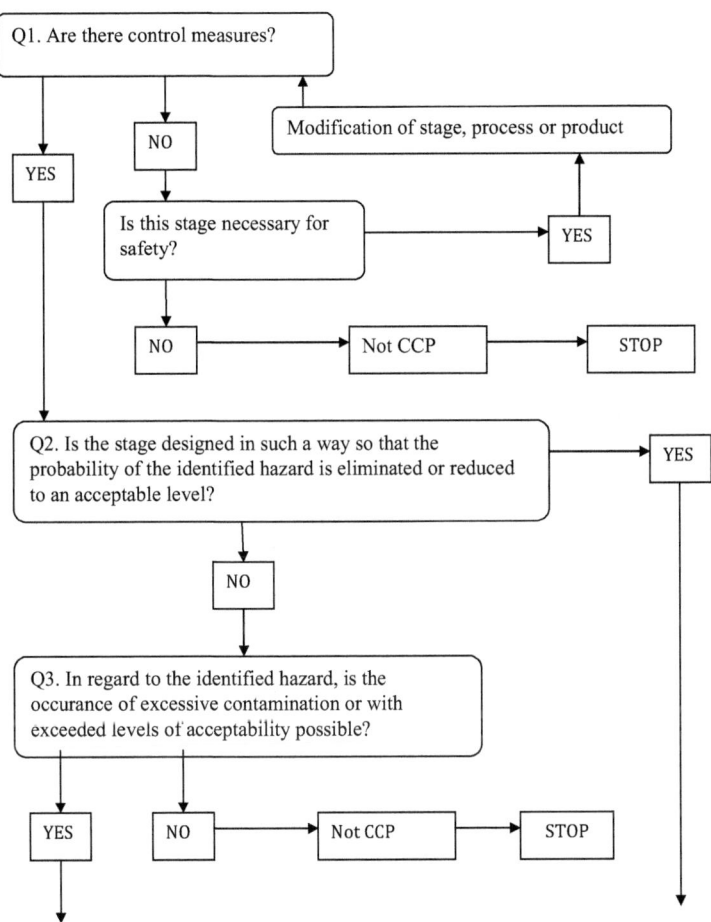

Q1. Are there control measures?

YES

NO

Modification of stage, process or product

Is this stage necessary for safety?

YES

NO → Not CCP → STOP

Q2. Is the stage designed in such a way so that the probability of the identified hazard is eliminated or reduced to an acceptable level?

YES

NO

Q3. In regard to the identified hazard, is the occurance of excessive contamination or with exceeded levels of acceptability possible?

YES

NO → Not CCP → STOP

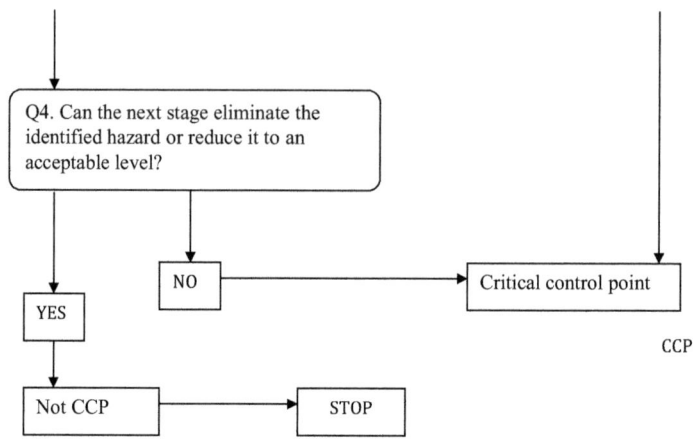

Figure no. 3.3. Based on the decision

Table 3.4 Identification of CCP in the Romano cheesemaking process

Product/Stage	Hazard	Decision tree questions				CCP/CP
		Q1	Q2	Q3	Q4	
Rennet	Biological	YES	NO	NO		CP
	Chimical	YES	NO	NO		
Pasteurisation	Biological	YES	YES			CCP
Cleaning	Biological	YES	NO	NO		CP
Coagulation	Chimical	YES	YES			CCP
Packaging	Biological	YES	NO	NO		CP
Storage Transport	Biological	YES	NO	NO		CP

30

Principle 3. Determination of critical limits

It is necessary to establish critical limits which must be respected in order to control each identified critical control point.

Critical limit is defined as the admitted tolerance for a certain parameter of critical control point. For a certain checkpoint, there may be one or more critical limits. If each of these limits has been exceeded, it means that this critical point is out of control and the innoucuity of the end product is threated.

The criteria used as critical limits may be: sensory parameters (appearance, color, taste, and smell), physical parameters (temperature, time, humidity), chemical parameters (acidity, salt content, pH), microbiological parameters such as yeast number, moulds, pathogenic bacteria. [1]

Table 3.5. Determination of critical points in the Romano cheesemaking process

CCP	Critical limits	Monitoring methods
PASTEURISATION	Temperature 63-65°C Time 15-20 minutes	- temperature and time control
COAGULATION	Acidity 20-21°T Temperature 38-40°C Time 15-18 minutes	- acidity, temperature, time control

Principle 4. Establishing a monitoring system that ensures effective control of critical control points

The monitoring process consists of the organized control of CCP and critical limits. The monitored results must be well documented and interpreted, as errors could lead to major product defects. All monitored results will be recorded, and the records and related documents will be signed by the persons who have conducted the monitoring process as well as by a management representative responsible for monitoring.

When monitoring, the operator must:

➢ position the measurement device;
➢ establish the measurement area;

31

> establish the measurement frequency;

> register the measured values;

> check through calibration with a reference device, at specified intervals, whether the measurement device is operating accurately. [1]

Table 3.6.Monitoring system in the Romano cheesemaking process

CCP	Risk	Critical limit	Monitoring				Corections and corrective actions	Responsable	Records
			What	How	Who	When			
Pasteurisation	Biological	Temp: 63-65°C Time: 15-20minutes	Temperature, Time	Visual	Pasteurisation operator	At each batch	Temperature and time adjustment	Pasteurisation operator	Monitoring sheet for pasteurisation temperature and time
Coagulation	Chimical Biological	Temp: 38-40°C Time 15-18 min	Temperature, Time	Visual	Coagulation operator	At each batch	Temperature and time adjustment	Coagulation operateur	Monitoring sheet for coagulation temperature and time

Principle 5. Establishing corrective actions that must be applied when the monitoring system indicates that a deviation from the specified critical limits has arrised

Corrective actions must be applied in order to eliminate existing risks or risks that may occur due to deviation from the HACCP plan, thus ensuring the finished product's innocuity. [1]

32

Table no. 3.7. Corrective actions in the Romano cheesemaking process

CCP	Monitoring	Corrective actions
Pasteurisation	Pasteurisation temperature and time monitoring by the operator.	Personnel training; Pasteurisation temperature and time adjustment
Coagulation	Coagulation temperature and time monitoring by the operator.	Personnel training; Coagulation temperature and time adjustment

Principle 6. Establish an efficient safekeeping system of the descriptive documentation (HACCP plan), the functional documentation (procedures and operational records related to the HACCP plan), representing the system documentation.

The HACCP plan needs to be available on a document format in the place where it will be applied to. Besides this plan, the documentation related to the CCP, the arrised deviations and corrective measures applied must be included. These documents will be made available to inspection authorities on request. [1]

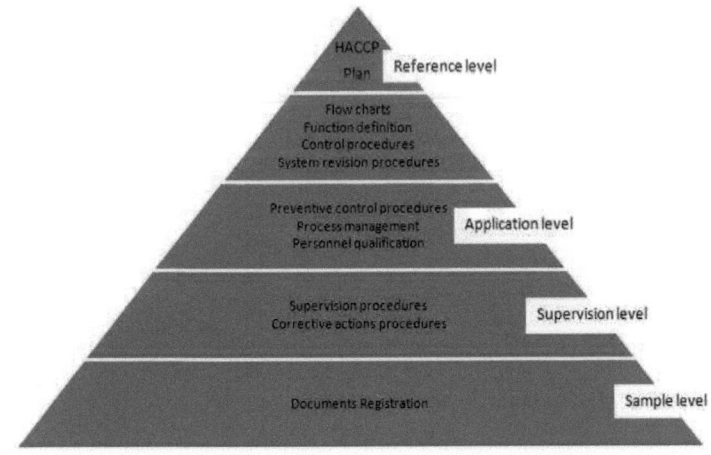

Figure no. 3.4. Documentary structure of the HACCP system

33

Principle 7. Establishing methods, procedures, specific tests for the functioning of the HACCP system, designed to:

- show compliance (if the HACCP system operates according to the HACCP plan)
- show effectiveness of the HACCP system (if the HACCP plan guarantees the security of the product).

Verification is made through methods, procedures and tests used to determine whether the existing HACCP system complies with the HACCP plan. Inspections can be made by the manufacturer, but also by control bodies. The inspection methods may be microbiological, physical, chemical and sensory.

Succesful implementation of the HACCP system is linked to pre-established good practice standards related to: building, equipment location, tehnological process, personell, cleaning and disinfection, raw and auxiliary materials used, product traceability, transport. [1]

Table no. 3.8. HACCP system verification

Function	HACCP audit-compliance				References
	System control questionnaire				Documentation Page:3 Date: 23.09.2015
	Name: Popescu First name: Ioan				
Procedure	Questions	C	NC	R	Observations
Evaluation procedure, raw material supplier selection, packaging materials	Was the reception of raw materials and supplier selection correctly performed?	X	-	-	-
Compliance with the repair and maintenance	Have the maintenance	X	-	-	-

34

procedure for the machinery, equipment and installations	responsables performed their tasks?				
Control of personnel health state and hygiene state for the work equipment	Is hygiene at work complied with?	X	-	-	-

4. TRACEABILITY STUDY OF ROMANO CHEESE MAKING

General aspects of traceability

A traceability system is a useful tool that helps a food chain organization achieve its defined objectives established within a management system.

Chosing a traceability system is influenced by regulations, product characteristics and beneficiaries expectations.

The complexity of the traceability system can very depending on the product features and objectives to be achieved.

Implementation of a traceability system within an organisation depends on:
- organization and product inherent technical limitations (ie nature of raw materials, batch size, collection and transport procedures, processing and packaging methods);
- the cost of applying such a system. [12]

Traceability systems should be able to document the history of the product and/ or locate a product on the food chain. Traceability systems contribute to research of non-compliance cuases and the ability to withdraw and / or recall products if necessary. Traceability systems can improve appropriate use and confidence in information, effectiveness and orgsanisation productivity.

Traceability systems should be:

- verifiable,
- applied consistently,
- result-oriented,

- efficient,

- with practical implementation.

In developing a food chain traceability system, it is necessary to identify the specific objectives to be achieved. Examples of objectives are:

- to support the food safety and/or quality objectives;
- to determine the product history or origin;
- to identify the responsible organizations within the food chain.
- to facilitate the verification of specific product information;
- to improve the efficiency, productivity and profitability of the organization. [12]

The traceability components are:

- Supplier traceability, represented by all records and documents on the basis of which the origin of all raw materials, ingredients, additives etc can be established;

- Process traceability, represented by records registered during the technological process, thus ensuring the possibility of identifying all raw materials, ingredients, additives from which a particular product was obtained as well as all operations these have undergone in the technological flow;

- Customer traceability, which ensures the identification of all customers of a certtain product. [8]

In order to make batch tracking possible, the traceability registrations must contain:

- the batch code;

- the work station code;

- date and time of registration.

The following critical events are to be registered, in order to avoid the loss of product traceability:

- Product creation
  - when a new batch code is given;
  - combining different batches;
  - when a batch is open;
  - when different data of different batches are added without losing their identity
- Reception
- Delivery

- Sale
- Product consumption [15]

Usefulness of traceability systems

Traceability systems are of great interest to consumers, food producers and processors, and to the state executive power.

Traceability systems are useful for consumers because:

- Specific foods and food ingredients that cause allergies, intolerance, or those who do not fit a certain lifestyle can be easily avoided
- the choice between different foods manufactured in different ways is possible.
- protection of food safety is possible by recognizing the product, if necessary

Therefore, a traceability system allows consumers to buy only secure food in terms of sanitation. [1]

Traceability system features

Basic characteristics of traceability systems implies identification and information and the link between them, these features being common to all systems, regardless of product, production and control system. Both products and processes are key components of the traceability system and, and therewith, may be essential and helpful information, as shown in the figure below.

Base components **Essential information** **Helpful information**

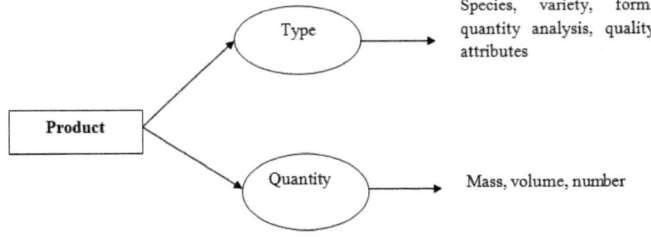

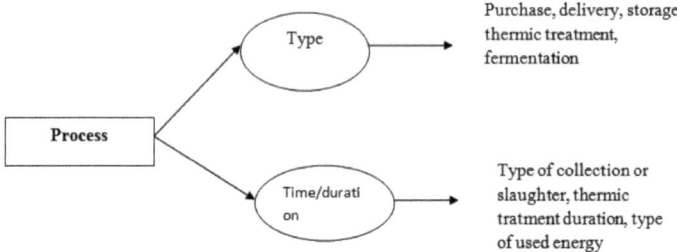

Figure no. 4.1. Traceability system features

[1]

Usage of traceability systems in the Romano cheese making process

In order to maintain production discipline and control, the implementation of an adequate sytem of identification and traceability is necessary.

Creating the necessary database to achieve traceability was carried out using the SQLite Manager program.

SQLite Manager is a relational database management system. This program is based on standard SQL language. The program offers multiple editions with different features that can satisfy a large variety of user requirements. [17]

SQLite is different from the majority of other SQL database engines in that it was designed to be:

- Easily managed.
- Easy to use.
- Easily encapsulated into a larger program.
- Easy to set and maintain. [16]

This program was created for database management within enterprises. [18]

To achieve the implementation of a traceability system to obtain Romano cheese, we created and used the following tables:

38

Figure no. 4.2. Tables

- **Batch**

We define the batch as the quantity of identical items that undergo changes along the technological process. Depending on the processing stage, various groups can be obtained, such as: raw material batch, auxiliary material batch, intermediate and finished product batch. A batch representing one day's production is taken as example.

| TABLE | lot | | | | Search | Show All | | | Add | | Duplicat |

rowid	cod	material_cod	dimensiune	timp
1	L01	01c	38812,88	2014-05-01 11:15
2	L02	04c	37903,92	2014-05-01 11:24
3	SM01	05c	715,16	2014-05-01 11:24
4	L03	06c	37790,21	2014-05-01 11:35
5	L04	08c	37383,19	2014-05-01 12:25
6	CO01	10c	7156,04	2014-05-01 13:23
7	Z01	11c	29816,83	2014-05-01 13:23
8	CO02	12c	7134,57	2014-05-01 13:50
9	B01	13c	7674,52	2014-10-01 15:00
10	B02	16c	7506,26	2014-10-01 15:12
11	CH01	02b	0,508	2014-05-01 12:00
12	CH02	02b	15,24	2014-05-01 8:00
13	CACL01	04b	7,26	2014-05-01 12:00
14	CACL02	04b	217,8	2014-05-01 7:30
15	DVS01	03b	0,53	2014-05-01 12:00
16	DVS02	03b	15,9	2014-05-01 7:45
17	SA01	01b	570,77	2014-05-01 13:51
18	SA02	01b	17123,1	2014-05-01 8:30
19	AMB01	01e	1,89	2014-10-01 15:10
20	AMB02	01e	56,7	2014-10-01 9:00

- **Workstation-** represents the process area every batch product goes through, starting with the raw material and finishing with the end product.

40

Table no.4.2. Workstation

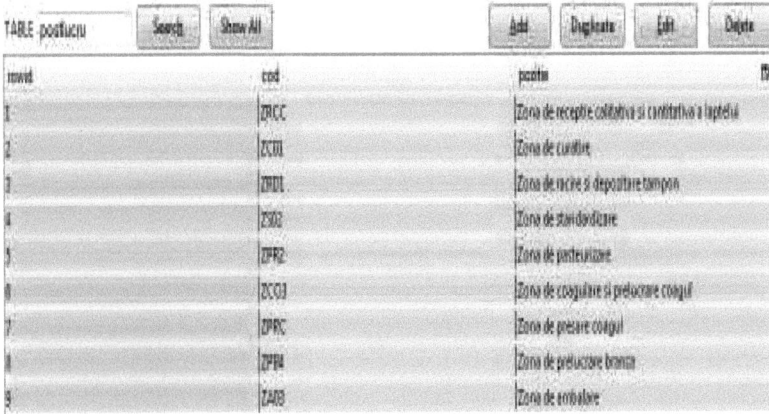

- **Batch_batch**

With the help of this table, a link between batches is established; pointing out that each batch is derived from the previous batch.

Table no.4.3. Batch_batch

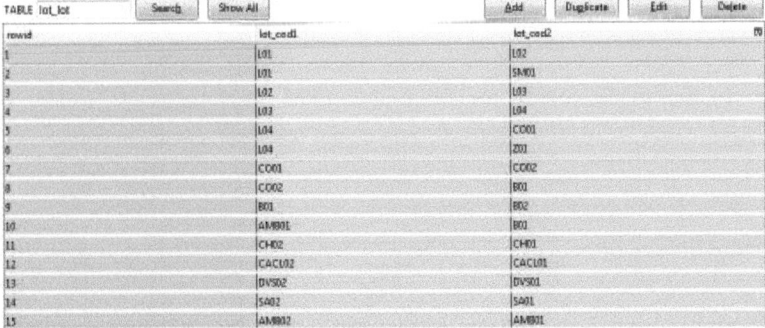

- **Traceability**

41

With the help of this table, every product batch can be monitored, thus obtaining information on: the workstation and the changes that batch has undergone, working parameters, responsible personnel for each workstation and duration of time in which each batch has been completed.

Table no.4.4. Traceability

TABLE Trasabilitate	Search	Show All		Add	Duplicate	Edit	Delete
rowid	lot_cod	postlucru_cod	in		out		▣
1	L01	ZRCC	2014-05-01 7:00		2014-05-01 9:31		
2	L01	ZC01	2014-05-01 7:15		2014-05-01 11:01		
3	L01	ZRO1	2014-05-01 7:30		2014-05-01 11:15		
4	L02	Z502	2014-05-01 7:40		2014-05-01 11:24		
5	L03	ZPR2	2014-05-01 7:50		2014-05-01 11:36		
6	L04	ZC03	2014-05-01 12:00		2014-05-01 12:25		
7	C001	ZC03	2014-05-01 12:30		2014-05-01 13:23		
8	C002	ZPRC	2014-05-01 12:50		2014-05-01 13:50		
9	B01	ZPB4	2014-05-01 13:51		2014-10-01 15:00		
10	B02	ZA03	2014-10-01 15:10		2014-10-01 15:12		
11	CH02	ZC03	2014-05-01 12:00		2014-05-01 12:06		
12	CACL01	ZC03	2014-05-01 12:00		2014-05-01 12:06		
13	DV501	ZC03	2014-05-01 12:00		2014-05-01 12:06		
14	SA01	ZPB4	2014-05-01 13:51		2014-08-01 13:51		
15	AMB01	ZA03	2014-10-01 15:10		2014-10-01 15:12		

To complete the necessary information, other related factory data was included in the database, such as: necessary materials, implemented operations during the technological process, machinery required for carrying out these operations, operating parameters, responsable persononnel, units of measurement for raw and auxiliary materials, personnel qualification.

- **Materials** - This table includes raw materials, auxiliary materials, intermediate products, finished product or their characteristics.

42

Table no.4.5. Materials

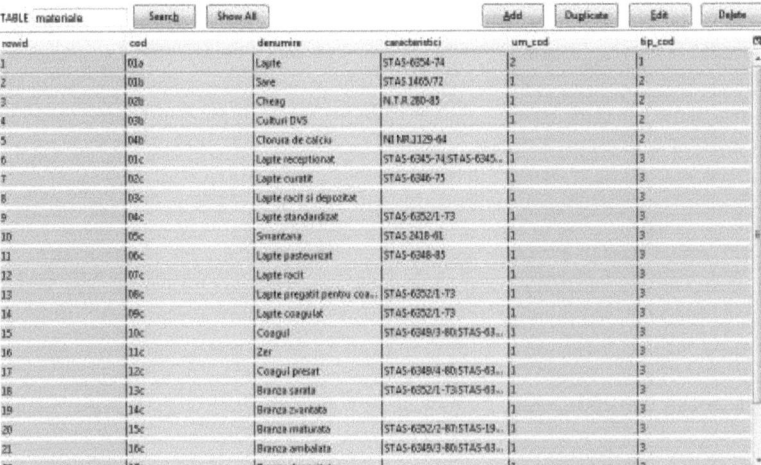

rowid	cod	denumire	caracteristici	um_cod	tip_cod	
1	01a	Lapte	STAS-6354-74	2	1	
2	01b	Sare	STAS 1465/72	1	2	
3	02b	Cheag	N.T.R 280-85	1	2	
4	03b	Culturi DVS		1	2	
5	04b	Clorura de calciu	NI NR.1129-64	1	2	
6	01c	Lapte receptionat	STAS-6345-74;STAS-6345..	1	3	
7	02c	Lapte curata	STAS-6346-75	1	3	
8	03c	Lapte racit si depozitat		1	3	
9	04c	Lapte standardizat	STAS-6352/1-T3	1	3	
10	05c	Smantana	STAS-2418-61	1	3	
11	06c	Lapte pasteurizat	STAS-6348-85	1	3	
12	07c	Lapte racit		1	3	
13	08c	Lapte pregatit pentru coa..	STAS-6352/1-T3	1	3	
14	09c	Lapte coagulat	STAS-6352/1-T3	1	3	
15	10c	Coagul	STAS-6349/3-80;STAS-63..	1	3	
16	11c	Zer		1	3	
17	12c	Coagul presat	STAS-6348/4-80;STAS-63..	1	3	
18	13c	Branza sarata	STAS-6352/1-T3;STAS-63..	1	3	
19	14c	Branza zvantata		1	3	
20	15c	Branza maturata	STAS-6352/2-87;STAS-19..	1	3	
21	16c	Branza ambalata	STAS-6349/3-80;STAS-63..	1	3	
22	17c	Branza depozitata		1	3	

- **Equipment**- refers to the equipment used to obtain the finished product.

Table no.4.6. Equipment

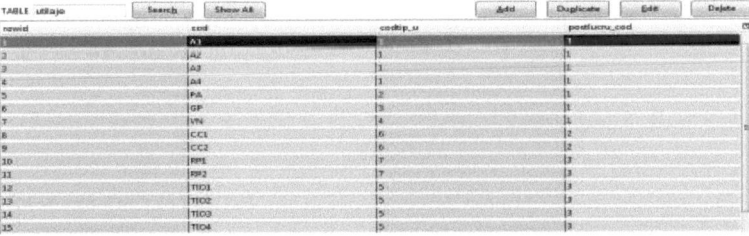

rowid	cod	codtip_u	postfucnu_cod	
1	A1		1	
2	A2	1	1	
3	A3	1	1	
4	A4	1	1	
5	PA	2	1	
6	GF	3	1	
7	VN	4	1	
8	CC1	6	2	
9	CC2	6	2	
10	PP1	7	3	
11	PP2	7	3	
12	TI01	5	3	
13	TI02	5	3	
14	TI03	5	3	
15	TI04	5	3	

- **Operation**- the respective table refers to the operations carried throughout the tehnological process, their description, duration and the exact workstation it is executed in.

43

Table no.4.7. Operation

cod	denumire	descriere	durata	postlucru_cod	
1	Receptie calitativa si cantitati...	Aceastã operatie are drept sc...	2,51	ZRCC	
2	Curatirea laptelui	Prin aceastã operatie se urmãre...	3,70	ZC00	
3	Racire si depozitare tampon	Operatia respectivã se aplicã at...	3,75	ZRD1	
4	Standardizare	Prin standardizare se întelege o...	3,73	ZS02	
5	Pasteurizare	Prin pasteurizare se asigurã dist...	1,67	ZPR2	
6	Racire	Aceastã operatie are drept scop...	3,06	ZPR2	
7	Pregatirea laptelui pentru coag...	Se realizeazã în urmãtoarele eta...	0,1	ZCO3	
8	Coagulare	Coagularea sau închegarea lapt...	0,3	ZCO3	
9	Prelucrarea coagulului	Dupã închegarea laptelui, coag...	0,9	ZCO3	
10	Presare coagul	Dupã introducerea în forme, co...	1	ZPRC	
11	Sarare coagul	Operatiunea de sãrare a brânzet...	21,00	ZPB4	
12	Zvantare	Zvântarea are loc în coloane de...	70	ZPB4	
13	Maturare	Imediat dupã sãrare, brânza cru...	20,00	ZPB4	
14	Ambalare	Ambalarea rotilor se face în hâr...	0,03	ZA03	

- **Parameter-** a table containing those parameters that are controlled during the technological flow.

Table no.4.8. Parameter

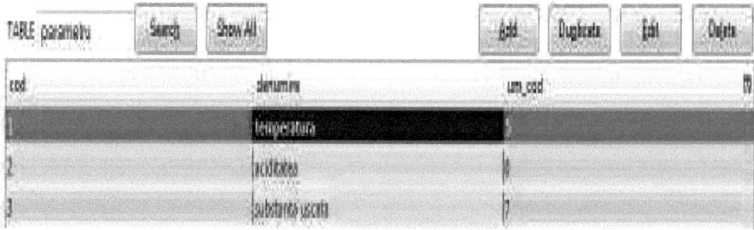

cod	denumire	um_cod	
1	temperatura		
2	aciditatea	0	
3	substanta uscata	7	

- **Personnel-** in this table is mentioned the factory's personnel, information regarding their contact details and their qualification.

44

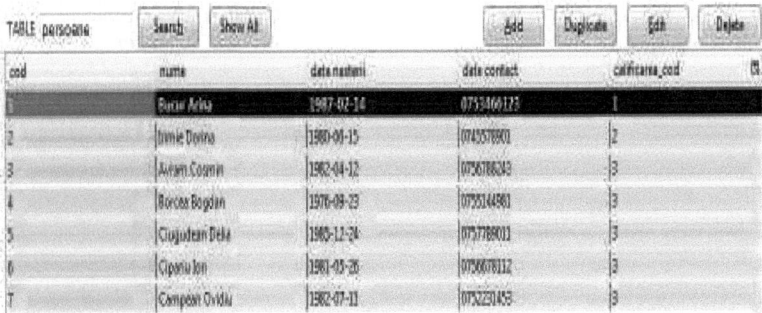

- **Um-** refers to the unit measurement equivalent to each product, parameter and their symbols.

Table no.4.10. Um

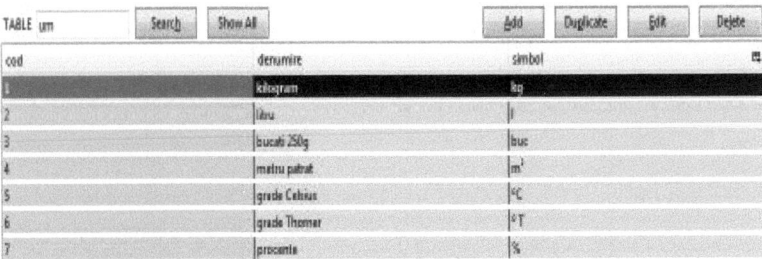

- **Qualification-** indicates personnel qualification.

Table no.4.11. Qualification

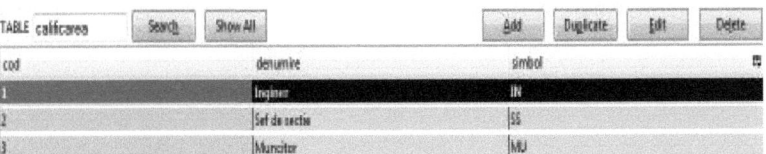

- **Type**

Table no.4.12. Type

Table no.4.12. Type

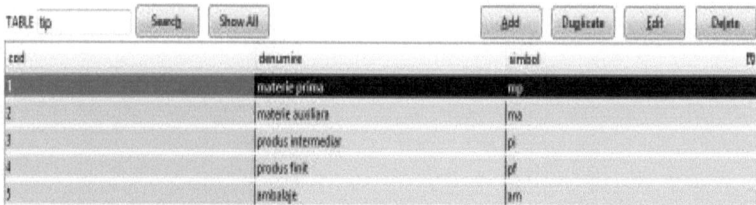

- **Type_u-** this table describes the machinery characteristics

Table no.4.13. Typ_u

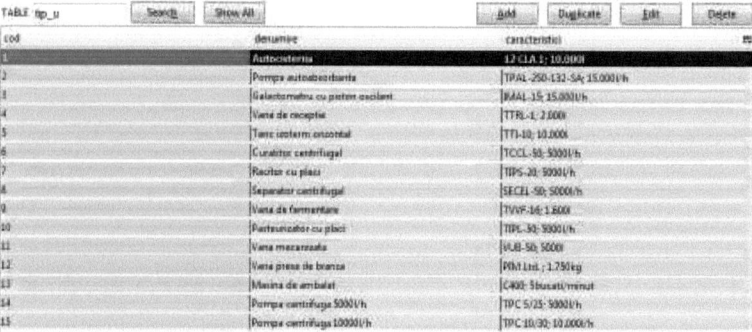

- **Records-** in this table are registered parameter values that correspond to a certain operation within a timeframe.

Table no.4.14. Records

TABLE inregistrari					Search	Show All				Add	Duplicate	Edit	Delete
rowid	operatie_parametru_cod	valoarea	timestamp	
1	1	9,2	2014-05-01 7:10	
2	1	9,3	2014-05-01 7:25	
3	5	64	2014-05-01 8:30	
4	5	64,2	2014-05-01 9:30	
5	5	64,5	2014-05-01 10:30	
6	8	39	2014-05-01 12:15	
7	8	39,5	2014-05-01 12:20	
8	8	39,5	2014-05-01 12:10	
9	8	39,7	2014-05-01 12:17	

- **Operation_parameter**- it is a link table, making a connection between the operation table and the parameter.

Table no.4.15. Operation_parameter

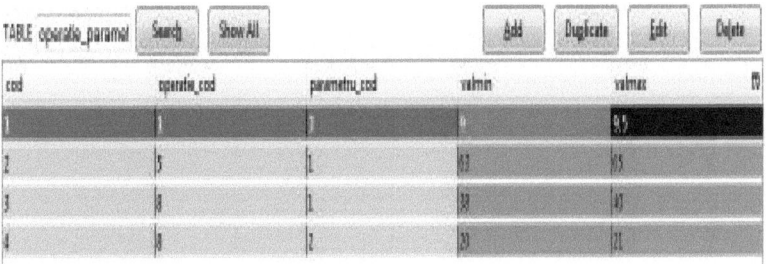

- **Operations_workstation**- makes the link between the operation table and the workstation table. Thus, it is easy to determine in which workstation each operation takes place, as well as its' duration.

Table no.4.16. Operation_workstation

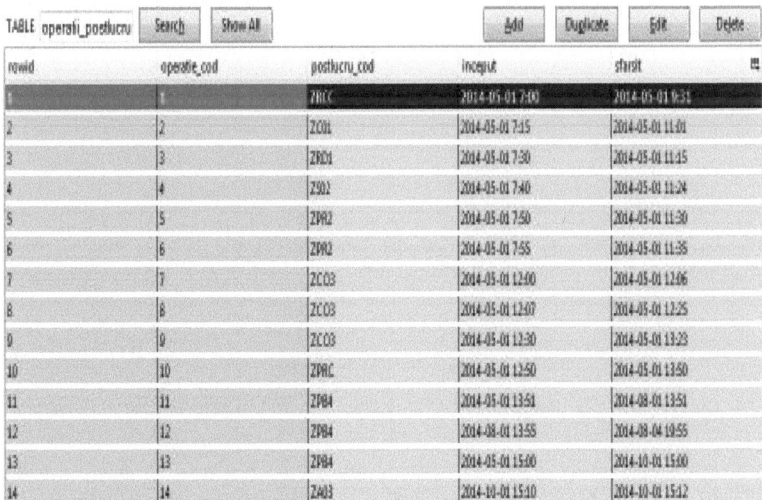

- **Personnel_workstation** - makes the link between the personnel table and the workstation. With the help of this connection table it can be determined which person is responsible of which workstation.

Table no.4.17. Personnel_workstation

rowid	persoane_cod	penflucru_cod	timp_in	timp_out	
1		ZRD1	2014-05-01 8:00	2014-05-01 8:30	
2	1	ZPB4	2014-05-07 10:00	2014-05-07 11:30	
3	2	ZRD1	2014-05-01 9:40	2014-05-01 9:50	
4	3	ZRCC	2014-05-01 7:00	2014-05-01 8:00	
5	3	ZRCC	2014-05-01 8:30	2014-05-01 9:31	
6	3	ZC01	2014-05-01 9:41	2014-05-01 10:41	
7	3	ZC03	2014-05-01 12:00	2014-05-01 12:06	
8	4	ZRCC	2014-05-01 8:00	2014-05-01 8:30	
9	4	ZC01	2014-05-01 7:15	2014-05-01 8:00	
10	4	ZC01	2014-05-01 10:45	2014-05-01 11:01	
11	4	ZRD1	2014-05-01 8:35	2014-05-01 9:40	
12	4	ZC03	2014-05-01 12:07	2014-05-01 12:25	
13	5	ZC01	2014-05-01 8:00	2014-05-01 9:40	
14	5	ZRD1	2014-05-01 7:30	2014-05-01 8:00	
15	5	ZRD1	2014-05-01 9:51	2014-05-01 11:15	
16	5	ZC03	2014-05-01 12:30	2014-05-01 13:23	
17	5	ZA03	2014-10-01 13:10	2014-10-01 13:12	

Database operation

For the batch of milk received the following should be identified:

- the workstations the batch has gone through,
- technological operations that are performed on that batch,
- personnel who had contact with this batch,
- batch size and the parameters to be checked for that specific batch.

SELECT operation.name, materials.name, personnel.name, batch.dimension, workstation. position, operations_workstation.start, operations_workstation.end, parameter.name, records. value, um.symbol, records.timestamp

FROM operation, materials, personnel, batch, workstation, operations_ workstation, parameter, records, operation_parameter, um

WHERE (operation.workstation _code= workstation.cod) AND (material_code=materials.code) AND

48

(personnel_workstation.personnel_code=personnel.code)   AND   (personnel_
workstation. workstation _code=

workstation.code)AND (batch.code LIKE "L01") AND (operations_workstation.
workstation   _code=

workstation.code)AND(operatii_postlucru.operatie_code=operatie.code)   AND
(operation_parameter.operation_

code=operation.code) AND (parameter.code=operation_parameter.parameter_code)
AND (

records.operation_parameter_code=operation_parameter.code) AND (value<valmin)
AND   (value>valmax)AND(parameter.um_code=um.code)   AND
(records.timestamp>="2014-05-01 7:10") AND (records.timestamp<="2014-05-01
7:25)

Results and discussions

postlucru_cod	denumire	persoane_cod	denumire	operatie_cod	denumire	parametru_cod	denumire	valoarea
ZRCC	Lapte	3	Bucur Arina	1	Receptie calitativa si cantitativa	3	substanta uscata	9,2
ZC01		1	Avram Cosmin	2	Curatire			9,3
ZRD1		4	Borcea Bogdan	3	Racire si depozitare tampon			

As resulted from the above table, the batch of milk has gone through three tehnological workstations, namely: quantity and quality reception area, cleaning area, cooling and storage area.

The technological operations performed on this specific batch of milk are: quality and quantity reception, cleaning, cooling and buffer milk storage. Also, three people working on those workstations were identified.

Note that the marked parameter for milk is the dry matter and the values recorded for this are 9.2, 9.3.

Conclusions

As noted in the above table, through database operation it can be determined the exact stage for each product batch used in the production of hard paste Romano cheese. Traceability represents the ability of an organization to retrace its food processing history within the production chain.

BIBLIOGRAPHY

1. Banu, Constantin, 2007, Suveranitate, securitate şi siguranţă alimentară, Editura ASAB, Bucureşti
2. Banu, Constantin, Vizireanu, Camelia, 1998, Procesarea industrială a laptelui, Editura Tehnică, Bucureşti
3. Chintescu, George, Toma, Alexandrina, Fabricarea brânzeturilor, Editura Tehnică, Bucureşti
4. Chintescu, George, 1980, Îndrumător pentru tehnologia brânzeturilor, Editura Tehnică, Bucureşti
5. Chintescu, George, Pătraşcu, Constantin, 1998, Agendă pentru industria laptelui, Editura Tehnică, Bucureşti
6. Costin, G.M., 2003, Ştiinţa şi Ingineria fabricării brânzeturilor, Editura Academica, Galaţi
7. Harbutt, Juliet, World cheese book, October 2008
8. Păduraru, Gabriela, 2010, Managementul siguranţei alimentelor
9. Răducuţă, Ion, 2004, Filiera laptelui, Editura Universităţii Lucian Blaga, Sibiu
10. Tiţa, Mihaela-Adriana, 2005, Tehnologii şi utilaje în industria laptelui şi a produselor din lapte, vol.II, Editura Universităţii Lucian Blaga, Sibiu
11. ***, Standarde şi condiţii tehnice de calitate
12. ***, 2007, SR EN ISO 22005:2007 – Trasabilitatea în lanţul alimentar, Principii generale şi cerinţe fundamentale pentru proiectarea şi implementarea sistemului, Bucureşti
13. http://scoaladebucatarieitaliana.blogspot.ro/2012/04/brinza-pecorino-romano-dop.html
14. http://www.gazetaromaneasca.com/timp-liber/gastronomie/1145-branzeturile-italiei-2-pecorino-romano.html
15. http://www.ift.org/knowledge-center/focus-areas/food-safety-and-defense/traceability.aspx
16. https://ro.wikipedia.org/wiki/SQLite
17. *http://www.techopedia.com/definition/1243/sql-server.* (2014)

18. *http://searchsqlserver.techtarget.com/definition/SQL-Server.* (2014)

51

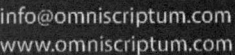